A Christian Vision of Science and Technology in Teilhard de Chardin

Agustín Udías, S.J.

En Route Books and Media, LLC
Saint Louis, MO

Make the time

En Route Books and Media, LLC
5705 Rhodes Avenue
St. Louis, MO 63109

Cover credit: Sebastian Mahfood with an image of Pope John XXIII and Pierre Teilhard de Chardin, stained glass window by Sieger Koder at Holy Spirit church in Ellwangen, Germany

Copyright © 2024 Agustín Udías, S.J.

ISBN-13: 979-8-88870-184-3

No part of this book may be reproduced, stored in a retrieval system, or transmitted in any form, or by any means, electronic, mechanical, photocopying, or otherwise, without the prior written permission of the author.

Table of Contents

Introduction 1

Chapter One: Science and Christian Faith 5

Chapter Two: Technology and Human Evolution 21

Chapter Three: Christological Dimension of Science

    and Technology 27

Conclusion 31

Introduction

Today, there is no doubt about the growing influence of science and technology on human life, which largely marks its progress. Although this has been true throughout history, it has increased markedly in the last hundred years.

In our time, more than ever, it is science that provides us with knowledge of the universe and of man's place and role in it. Science presents us with the nature and structure of matter and of the universe, and especially of living beings, including mankind. It presents us with an immense, though finite, universe, made up of thousands of millions of galaxies, each with millions of stars around some of which revolve planets, one of which is our Earth.

This universe has evolved from the beginning, some fourteen billion years ago, from what we call the "big bang." Only elementary particles of atoms existed at the beginning, and molecules of increasing complexity formed stars and planets over time. Life may have been able to develop on many planets, and intelligent beings evolved on, at least, our Earth.

Science, which has provided us with such a vision of the universe, has today become a worldwide phenomenon. Today, it is estimated that the number of scientists has increased enormously and is about nine million, spread over all countries. Naturally, the richer countries have a greater share. The most developed countries, known as the G20, account for 89% of all researchers and produce 91% of all scientific publications. Developing countries, however, are also increasing their share.

Technology, with the practical applications of science, has made possible today unanticipated levels of human welfare, which are becoming available to the entire world population. Science and technology are sometimes considered as a single phenomenon under the name of "technoscience." To recognize their importance, one need only acknowledge the many advances in medicine, in land, sea, air and space transportation, and in new developments in computing, social networking, artificial intelligence, and genetic engineering, to name but a few.

Science and technology, although in different proportions, have been spreading rapidly throughout all countries, not only the most developed ones, contributing to the process of globalization that is unifying the different lands, nations, and races in what has already been called a "global village" as Marshall McLuhan coined in 1962.[1]

---

[1] Marshall McLuhan and Bruce Power, *The Global Village: Transformations in the World Life and Media in the 21st Century* (Oxford: Oxford University Press, 1992).

Chapter One

Science and Christian Faith

The problem posed by the relationship between science, technology, and the Christian faith has undergone a profound change in recent years. It is no longer the problem of certain difficulties that may exist between some statements of science and theology, but a general questioning that is more subtle and difficult to analyze.

At present, the danger is not in concrete difficulties, but in a general ideology and a widespread presupposition that often accompany the world of science and technology. The natural sciences that today provide the knowledge of the world proclaim themselves, without formulating it scientifically, as the living force that marks the progress of the world. This ideology is present in a large part of the people active in the different fields of natural sciences and technology who collaborate in the construction of the world. Scientists and engineers, by their great achievements in understanding and applying the laws of nature, have developed a self-consciousness in their methods and accomplishments that leads them to consider themselves and their work as prevailing over others.

The Second Vatican Council, in its pastoral constitution "The Church in the Modern World" (*Gaudium et Spes*) draws attention to this position:

> Indeed, today's progress in science and technology can foster a certain exclusive emphasis on observable data, and an agnosticism about everything else. For the methods of investigation which these sciences use can be wrongly considered as the supreme rule of seeking the whole truth. By virtue of their methods these sciences cannot penetrate to the intimate notion of things. Indeed the danger is present that man, confiding too much in the discoveries of today, may think that he is sufficient unto himself and no longer seek the higher things.[1]

Thus, the Church recognizes this problem that touches for many the very roots of faith. It is a much

---

[1] The Church in the Modern World" (*Gaudium et Spes*), 57. www.vatican.va/archive/hist_councils/ii_vatican_council/documents/vat-ii_const_19651207_gaudium-et-spes_ en.html.

more important problem than the concrete difficulties because it creates in modern man a spiritual structure that can conflict with the Christian faith itself.

The sciences demand total submission and present the security that comes from being grounded in their observational and mathematical method, which guarantees that phenomena are treated in an objective and verifiable way. The sciences provide an understanding of the observable world through a complex process of observation, hypothesis, and verification, always open to revision when new empirical data so require. Christian faith, on the other hand, moves at another level, namely the recognition of God's salvation in Jesus Christ. Science can also create a kind of faith that often results in a spiritual dualism, in which its scientific understanding of the material world can separate it from faith in God.

The consequences of this stance can contribute in Christians, especially those engaged in scientific and technological work, to a lack of personal synthesis between faith and science that can eventually lead to a rupture in the unity of their inner life. One aspect of their life is devoted to scientific work, and another is reserved for spiritual religious activities and interests. As a result, the problem itself may become ignored. We can pretend that there is no conflict between the scientific and religious positions as both can exist side by side, one close to the other in the same person without any influence of one on the other.

We can see, however, that this is not a good solution since the unity and harmony of the personal attitude is somehow broken and a kind of spiritual schizophrenia may result. Seeking to keep these two forces of our spiritual life isolated would mean losing much of their strength and the light that their harmony and integration should bring. On the one hand, we must bring by faith the possibility of integrating into the spiritual life the positive achievements of the field of scientific work and, on the other hand, draw from scientific work the strength and inspiration that will illuminate by faith the future of men and of the universe. We must also bring with our faith, the testimony of a spiritual sense for a world built by means of technology. Only when our religious position, in some way, incorporates the scientific and technical position can its testimony have an influence on today's technological world. Today's world is undoubtedly structured and constructed through the influence of scientific thought and the achievements of technological progress. The Council clearly recognizes this when it says:

> Today's spiritual agitation and the changing conditions of life are part of a broader and deeper revolution. As a result of the latter, intellectual formation is ever increasingly based on the mathematical and natural sciences and on those dealing with man himself, while in the practical order the technology which stems from these sciences takes on mounting importance. This scientific spirit has a new kind of impact on the cultural sphere and on modes of thought. Technology is now transforming the face of the earth, and is already trying to master outer space.[2]

It cannot be denied today that it is technology and science that are structuring and modifying culture and thought in a broad way. Consequently, the

---

[2] The Church in the Modern World (*Gaudium et Spes*), 5.

witness of our faith in this world must participate in all its efforts and concerns. Our faith itself must participate in the efforts and achievements of world building through science and technology.

To achieve this, we need to acquire for ourselves a harmony between these two human aspirations. To arrive at this harmony, we need to deepen, on the one hand, the strength and possibilities of science and technology and, on the other hand, the central aspect of our Christian faith. We experience step by step today a continuous discovery of the growing importance of science and technology in the world. We should not be afraid to recognize their importance, nor should we be afraid to try to impose *a priori* limits on how far they can go. We have to admit that science today has changed to a large extent even human consciousness itself. Faced with this situation, we can ask ourselves if, from the Christian faith, we can find a positive sense in the scientific and technological phenomenon, which, from a religious point of view, many view with a certain caution and fear for promoting a materialistic vision of the world.

Pierre Teilhard de Chardin (1881-1955), Jesuit geologist and paleontologist, raised this problem

Chapter One: Science and Christian Faith

many times from his Christian and scientific point of view.[3] First of all, Teilhard was aware of the importance of the role of science today in the world, as he affirms: "After a century, in the world, scientific research has become, both quantitatively (for the number of persons that are engaged) and qualitatively (for the importance of the results obtained), one of the greatest forms but the main form of the reflective activity in the earth."[4] Thus, for Teilhard, scientific research is not a part of human endeavor, important as it is, but, as he put it in an essay with the explicit title "The Religious Value of Scientific Research," it constitutes the "great business of the world" (*la Grande Affair du Monde*), "the vital human

---

[3] Robert Speaight, *Teilhard de Chardin. A Biography* (London : Collins, 1967), Claude Cuénot, *Pierre Teilhard de Chardin. Les grandes étapes de son évolution* (Paris: Plon, 1958).

[4] The English text of Teilhard's quotations is my translation from the original French text. References are given to the French edition of Teilhard's works: *Œuvres de Pierre Teilhard de Chardin* (Paris: Édition de Seuil, 1955-1976). "Recherche, travail et adoration," *Œuvres*, 9, 284.

function, as vital as nutrition and reproduction."[5] The importance of science for the development of human life can hardly be expressed more emphatically. If Teilhard said this in 1947, today, seventy-six years later, after enormous scientific progress, it is even more true.

Teilhard posed the problem of the relationship between science and the Christian faith explicitly in 1921 in a lecture entitled "Science and Christ or analysis and synthesis,"[6] where he addressed his listeners, as he himself says: "to make them love science in a Christian way," discarding any attitude of distrust and fear, as if science were the enemy of faith. In this lecture, Teilhard began by recognizing the limits of scientific analysis, which seeks above all to find the simplest constitutive elements of things and of the world. This is, according to him, "necessary and good, but it cannot lead us where we are interested," especially in religious consideration, where a synthesis

---

[5] "Sur la valeur religieuse de la recherche," Œuvres 9, 258.

[6] "Science et Christ ou analyse et synthèse," Œuvres, 9, 45-62.

approach is necessary to find even a true meaning to science itself.

In this way, Teilhard continues, those who think that "science is so strong that it alone can save us"[7] are deceiving themselves—a wake-up call for those who want to find in science a substitute for religion. Therefore, accepting all the knowledge that science is providing us with about the world, "science itself must not disturb us in our faith with its analyses, but, on the contrary, must help us to know, understand and appreciate God better." Teilhard, finally, concludes, in a way that may astonish many with spiritualistic tendencies so common today, "I am convinced that there is no more powerful natural nourishment for the religious life than contact with well-understood scientific realities."[8] It is worthwhile to take this seriously and stop seeing science as something completely unrelated to any religious consideration.

This could be said of any kind of religion, but Teilhard goes a step further by considering the special relationship between scientific knowledge and

---

[7] Ibid., 54, 57.
[8] Ibid., 61-62.

Christian faith. For Christianity, God is not only the creator, but has also become incarnate in the world in Christ, and thus Teilhard can affirm that God, "by his incarnation, is interior to the world, is rooted in the world down to the heart of the smallest atom." For the Christian, therefore, by his incarnation, God has united himself in Christ to the material universe which, through science, we know to be dynamically evolutionary.

Therefore, Teilhard can conclude, "It is unjust to oppose science and Christ or to separate them as two domains foreign to each other."[9] So Teilhard can still take a step forward, and in the above-mentioned essay, he concludes: "because scientific research (followed with faith) is the only ground on which the human-Christian mysticism can be elaborated which can create tomorrow a true human unanimity."[10] This may be difficult for us to understand, but it must be kept in mind that Teilhard places himself in a new Christian mysticism, as will be seen below.

---

[9] Ibid., 62.

[10] "Sur la valeur religieuse de la recherche," Œuvres 9, 263.

In particular, referring to the evolutionary vision of life and the universe, which science presents to us today, and which is sometimes considered opposed to the Christian faith, Teilhard in another writing affirms: "Christianity and evolution are not two irreconcilable visions, but two perspectives that fit together and complement each other."[11] The figure of Christ, under the invocation so dear to Teilhard of the "cosmic Christ," allows him to affirm: "The great cosmic attributes of Christ (especially present in St. Paul and St. John) are those that give him a universal and final primacy over creation;[12] to finally establish: "Evolution is the daughter of science, but after all, it may well be faith in Christ that will save our appreciation of evolution tomorrow."[13]

For Teilhard, science itself must be understood in the context of evolution as one of its essential elements: "Scientific research is the very expression (on the plane of reflection) of the evolutionary effort, not only to subsist but to become more, not only to

---

[11] "Catholicisme et science," *Œuvres*, 9), 240.
[12] Ibid., 239.
[13] Ibid., 240, 241.

survive but to survive irreversibly."[14] Considering the dynamic evolutionary nature of the world known to science, for Teilhard, the Christian mystery of the incarnation itself takes on a special significance since it must also be understood within the evolutionary process of the world.

In 1953, a visit to the cyclotron at Berkeley, California, led Teilhard to regard it as a symbol of scientific and technical progress, and thus, he sees in it: "A whole range of knowledge and techniques, a whole spectrum of energies, too, converging there where I am."[15] Faced with this symbol of the most advanced research of the time in the field of physics, Teilhard discovers a new meaning and a deeper dimension of scientific research:

> "To my eyes, what we simply call 'research' appeared charged, colored and enkindled with certain potentialities (faith, worship) hitherto considered as foreign to science.... Looking at it more

---

[14] "Sur la valeur religieuse de la recherche," Œuvres 9, 258.

[15] "En regardant un cyclotron," Œuvres, 7, 367-377, 369.

closely, I see this research, forced by an inner necessity, concentrating, in short, its efforts and hopes in the direction of a divine focus."[16]

Here we see how Teilhard discovers in scientific research itself an intrinsic religious value.

In one of his last essays, written in 1955, the same year of his death, with the title "Research, Work and Worship," Teilhard goes a step further and finds in scientific research a form of worship.[17] He begins by recognizing the importance of scientific research in the modern world, so evident today. He then adds that for a Christian:

"[R]eligiously speaking, the results and achievements of science can be considered as an accessory or an augmentation of the Kingdom of God, since finding ourselves in a convergent universe as revealed by science (and only in such a universe) Christ finds the fullness of his creative action, thanks to the existence, finally perceived, of

---

[16] Ibid., 376-377.
[17] "Recherche, travail et adoration," Œuvres, 9, 283-289.

a natural and supreme center of cosmogenesis where he can situate himself."

By convergent universe, Teilhard means one that by its evolution will finally converge in what he calls the "Omega Point," that is, a final transcendent culmination in God. His Christian faith identifies the Omega Point with Christ, so that the cosmogenesis of evolution becomes what he calls a true "Christogenesis."

The central point of the Christian faith is the Mystery of Jesus Christ, God incarnate in the world. We are accustomed to consider Christ as God made man, and we seldom consider its consequence that God is united to a part of the material universe. Just as in Christ there is a divinization of man, there is also a divinization of matter. Matter, which in man is the bearer of the spirit, is, beginning with the incarnation, the means of God's definitive revelation to the world.

In Jesus Christ, the convergent point of the whole process of evolution of the universe, the union of God with men and through it, the union with the whole material universe, really takes place. This is why Teilhard can say: "I greet you, Matter, divine

Medium, charged with creative power, ocean agitated by the Spirit, clay kneaded and animated by the incarnate Word."[18] Thus, the sanctification of man considered as the axis of the evolution of the universe is not something of the future, as we used to think, but by the power of God who becomes man in Christ is already present through faith.

In this regard, St. Paul says: "God has made known to us his hidden purpose, which will be accomplished when the right time comes: namely, that the universe, all that is in heaven and on earth, be brought to unity in Christ" (Eph. 1:9). Thus, Teilhard can finally conclude that, through Christian thought and prayer, for the one, whom he calls "the believer of tomorrow," scientific research becomes "a new and superior form of worship."[19]

---

[18] "La puissance spirituelle de la matière," *Œuvres* 12, 479.

[19] "Recherche, travail et adoration," *Œuvres*, 9, 289.

Chapter Two

Technology and Human Evolution

In addition to science, the practical application of technology is for Teilhard today a key element of human evolution. He developed this idea in his 1948 essay, "Place of Technology in a General Biology of Humanity."[1] He begins by recognizing the role of technology in the modern world: "Man has entered the age of industry with its aspect of socialization" and asks: What is the meaning of this important fact that inaugurates a new period? to answer: "Industrial progress is not something accidental, but constitutes an event capable of gathering the greatest spiritual consequences."[2] For Teilhard, technological progress is a fundamental part of the world evolutionary process at the human level, so he can say: "To understand the place of technology in human society it is necessary to go back to the general process of world evolution" and

---

[1] "Place de la technique dans une biologie générale de l'humanité," Œuvres 7, 161-169.
[2] Ibid., 161.

concludes: "Technology has a biological role; therefore, it belongs by its own right to the field of the natural."[3] This is important because technology is often considered on the level of the artificial and therefore outside and even contrary to the natural and ecological. Teilhard, on the contrary, considers technology precisely within the natural process of evolution as its expression at the human level.

Already in his fundamental work, *The Phenomenon of Man*, Teilhard considers the modern evolution of the "noosphere," a term he uses to designate the thinking envelope of the earth, just as the biosphere is its living envelope, and recognizes that we are currently in a "change of epoch," above all, due to technological progress that demands a change of thought:

> "The age of industry. The age of oil, electricity and the atom. The age of the machine. The age of great collectivities and science..... A smoking land of factories. A land accelerated by business. An earth vibrating with hundreds of new radiations. This great organism does not live defi-

---

[3] Ibid., 166.

nitively but by and for a new soul. The change of epoch is demanding a change of thought."[4]

Technological development thus constitutes an important element in the new dynamic and irreversible process at the planetary level, as part of cosmic evolution, which Teilhard calls "socialization" and "planification" at the human level, and which we also know today as "globalization." For him, cosmic evolution continues on earth at the human level of the Noosphere, as an increase in consciousness. This implies a progress of humanity toward a new, more evolved stage that Teilhard calls the "hyperpersonal" and the "ultrahuman." From this stage, humanity will eventually converge towards the Omega Point that his Christian faith identifies with Christ.

In this evolution, which Teilhard describes as "a human tide that irresistibly lifts us up... the relentless ascent on our horizon of a true Ultra-human," science and technology play an essential role. Thus, he joins the role played by science and technology in human evolution:

---

[4] "Le phénomène humain" *Œuvres* 1, 238.

> "The truly explosive development of technology and research, the mastery at once theoretical and practical over the secrets and resources of cosmic energy in all its degrees and under all its forms, leads correlatively to the rapid elevation of what we have called the psychic temperature of the Earth."[5]

With the "elevation of the psychic temperature of the Earth" and the "ultrahuman," Teilhard designates the most advanced stages of human evolution.

It has already been mentioned how the modern phenomenon of globalization, which we are beginning to experience, can be interpreted as a sign, even a weak one, of the human convergence postulated by Teilhard. To this sign we can add others, largely also consequences of scientific and technical development, which are emerging in human society and which can also be interpreted in this sense, for example, the increase in global communications, the growing concern for world problems, and the institutional strengthening of international

---

[5] "Sur l'existence probable, en avant de nous, d'un « ultra-humain," *Œuvres* 5, 359-360.

organizations (United Nations, International Court of Justice, etc.).

Modern times, however, also witness many divergent trends, such as nationalism, autocratic governments, terrorism, violence and wars. Technology, which has helped to create many conditions that foster human unity, is also responsible for negative aspects such as the arms industry and ways of life that encourage individualistic tendencies such as consumerism and growing social inequalities.

Faced with this situation, we can ask ourselves whether there are reasonable grounds for maintaining Teilhard's optimistic position. Today we need some of his optimism to be able to see, through the many dark signs, the light that shines in the distance as hope for the future of humanity and the development of what he called the Ultrahuman. It is precisely the Christian faith that can assure us of that future which will finally be reached through the union of men and everything in Christ, the true Omega point of the evolution of the universe.

Chapter Three

Christological Dimension of Science and Technology

Teilhard's Christian vision of science and technology finally has its foundation in the recognition of their importance in the human part of cosmic evolution which ends in the convergence at the Omega Point which is Christ, that is to say, the "Christological dimension of the universe."[1] Teilhard introduces the role of the Christian faith precisely in the context of the convergence of human evolution with which it is in consonance, and thus constitutes the "religion of the future," by implicitly recognizing what he calls the "human sense."[2] This human sense is what drives man to his consummation in a final unity and, for

---

[1] André Dupleix y Évelyne Maurice, *Christ présent et universel. La vision christologique de Teilhard de Chardin* (Paris : Mame-Desclée, 2008) ; François Euvé, *Por una espiritualidad del cosmos. Descubrir a Teilhard de Chardin* (Maliaño: Sal Terrae, 2023).

[2] "Le sense humain," *Œuvres,* 11, 21-44.

Christians, this will be achieved in the final union of men in Christ.

Teilhard concludes that Christ is the only one who can truly save the human aspirations of our time, in which science and technology play a crucial role. Therefore, he can say that "the light of Christ is not eclipsed by the brightness of the ideas of the future, of science and progress, but occupies precisely the center that holds its fire."[3] In this way, Teilhard proposes a Christian interpretation of the whole cosmic evolution that leads to an Omega Point that he identifies with Christ, God incarnated in the world.

In the human stage of evolution, the attraction to the Omega Point that drives the noosphere by the force of love toward its final convergence takes place through the historical presence of Jesus of Nazareth. In him, the presence in the noosphere of the ultimate center toward which it tends is realized. He is, therefore, the presence of the Omega Point in human history, drawing everything, even science and technology to himself by love, and in him everything will find its final consummation. In this way, Teilhard finally resolves the tension between the free nature of

---

[3] Ibid., 41.

man and his convergence towards unity. In Teilhard's interpretation, the cosmogenesis of evolution becomes a true "Christogenesis," since the definitive pole or center of evolution is identified with Christ, that is, God incarnate. The unity in Christ of the whole universe, including humanity, through, among other elements, science and technology, is what Teilhard calls the "Universal or Total Christ."[4]

Finally, the presence of Christ in the world leads Teilhard to consider the world itself, including in it all the progress of science and technology, as what he calls a "Christified world." Thus, he will say that the presence of Christ-Omega converts the cosmic dimension of the world into a "Christic" dimension, in such a way that the cosmic expands and enlarges the Christic, and the Christic "fills with love" (*s'amorise*, a term used by Teilhard to express the expansion of love). That is, it fills with energy (the energy of love) to the point of "incandescence" in the field of the cosmic.[5]

For Teilhard, therefore, what he calls the Christic constitutes a synthesis of cosmic convergence and

---

[4] "Le Christique," *Œuvres* 13, 93-118.
[5] Ibid., 110-113.

Christic emergence. He unites, in this way, the vision from below, where science and technology act in a special way, with a vision from above that comes from God. He unites what can be reached by contemplating the world in evolution (the cosmic dimension where science and technology enter) and what the Christian faith tells us of Christ (the Christic dimension) present in the world through his incarnation. Therefore, Teilhard can clearly affirm:

> "By virtue of Creation and above all of the Incarnation, nothing is profane, here on earth, for those who know how to see it (we can add neither science nor technology). On the contrary, everything is sacred, for those who distinguish in every creature, the trace of having been chosen and submitted to the attraction of Christ on the way of consummation."[6]

The negative visions of science and technology would, therefore, remain outside Teilhard's Christian vision of the world.

---

[6] "Le Milieu divin," Œuvres 4, 56.

Conclusion

Faced with the growing influence of science and technology in the modern world, often viewed with some suspicion and fear, especially from religious considerations, Teilhard presents us with a positive Christian vision of them. Science has uncovered for us the nature of an enormous evolving universe in which life and intelligence have evolved and in which technology has made human progress possible by increasing and improving the quality of life and extending it to the entire human population.

From the Christian point of view, this is the universe created by God and in which Christ has become incarnate and is present. According to Teilhard, evolution continues to follow the direction from matter to spirit and progresses at the human level, largely thanks to science and technology, to finally converge by its attraction on an Omega Point which is Christ himself. In this way, the cosmogenesis of evolution becomes a Christogenesis. Science and technology are, therefore, important elements in the process that Teilhard calls "christifying" the world toward its final convergence in Christ at the end of time.

www.ingramcontent.com/pod-product-compliance
Lightning Source LLC
Chambersburg PA
CBHW070047070426
42449CB00012BA/3184